Hydrogen Safety 2023

Topical Conference at the 2023 AIChE Spring Meeting and 19th Global Congress on Process Safety

Houston, Texas, USA
12-16 March 2023

ISBN: 978-1-7138-7936-7

Hydrogen Safety 2023

Topical Conference at the 2023 AIChE Spring Meeting and 19th Global Congress on Process Safety

Houston, Texas, USA
12-16 March 2023

AIChE
The Global Home of Chemical Engineers

Printed from e-media with permission by:

Curran Associates, Inc.
57 Morehouse Lane
Red Hook, NY 12571

proceedings.com

Some format issues inherent in the e-media version may also appear in this print version.

AIChE
120 Wall Street, FL 23
New York, NY 10005-4020

Phone: (800) 242-4363
Fax: (203) 775-5177

www.aiche.org

Additional copies of this publication are available from:

Curran Associates, Inc.
57 Morehouse Lane
Red Hook, NY 12571 USA
Phone: 845-758-0400
Fax: 845-758-2633
Email: curran@proceedings.com
Web: www.proceedings.com

TABLE OF CONTENTS

119a Consequence Modeling Case Study for Hydrogen Transportation and Storage

Tatiana Norman, Tatiana Flechas, Cynthia Spitzenberger

Considering the growing interest in using hydrogen as fuel, it is essential to be able to transport and store hydrogen under high pressure or as a liquid in large quantities safely.
Hydrogen transportation and storage pose safety challenges due to its wide flammable range, high buoyancy, low minimum ignition energy, and wide range of explosion consequences. Flammable dispersions, jet fires, pool fires, and explosion scenarios (including BLEVE, Boling Liquid Expanding Vapor Explosion) are some of the potentially hazardous consequences of a hydrogen release scenario.
DNV has modelled the consequences of hydrogen releases from the following sources: 1) High-pressure large diameter underground transmission line (approx.24"), 2) High-pressure above ground transmission line (approx. 12''), 3) High-pressure storage tank, and 4) Low-pressure bullet storing liquid hydrogen. The consequence analysis was conducted for each of the cases using DNV proprietary software Phast.
In this analysis, the release size was assumed to be 3% of the cross-sectional area of the pipe or storage cylinder, which corresponds to the NFPA 2 approach for determination of the recommended separation distances.
The magnitude of the potential hazard zones for each release event was assessed based on flammable gas dispersion (Lower Flammable Limits), jet fire radiation levels, and explosion overpressure. The setback distances were also calculated and compared with existing industry standards such as OSHA 1910.103 and NFPA 55.

119b Maximizing Smr Furnace Capacity without Compromising Tube Wall Integrity

Alina Green

Temperature measurement and monitoring is key to performance and safety, especially during the potentially dangerous startup process. As a result of exposure to the high temperature, the tubes inside a primary reformer furnace operate close to their metallurgy temperature limits.

This paper will discuss how tubeskin thermocouples can help improve IR scanning results in steam methane reformers as verified in the Wika R&D process unit. Differentiating between the radiation emitted by the target tube and the surrounding hot surfaces, temperature loss at the viewing port and the inability to reach target tube due to SMR geometry are just a few of the challenges operators face when using IR scanning for measuring tube temperature. Tube skin temperature estimation can be highly inaccurate without an accurate reference temperature for fine-tuning emissivity.

We will cover why continuous tube temperature monitoring is critical during startup and what is the cost of inaccuracy while we go over a case study about how tubeskin temperature monitoring can help operators reach furnace full capacity without compromising tube integrity.

Finally, we will discuss the results of the patent-pending new generation tubeskin thermocouples developed in Wika's R&D center with even higher accuracy and simplified one-step installation ideal for SMR furnace tubes.

SPRING23
+19TH GCPS
A Joint AIChE and CCPS Meeting

Blended Natural Gas / Hydrogen Fuel Gas Systems: An Evaluation of Risk

Nicholas Reding
Exponent, Inc.
10850 Richmond Ave., Ste 175, Houston, Texas 66045
nreding@exponent.com

Alfonso Ibarreta,[a] Achim Wechsung,[a] Ryan Hart,[b] Delmar "Trey" Morrison[c]
Exponent, Inc.
a – 1075 Worcester Street, Natick, Massachusetts 01760
b – 10850 Richmond Ave. #175, Houston, Texas 77042
c – 4580 Weaver Parkway, Suite 100, Warrenville, Illinois 60555
myen@exponent.com, awechsung@exponent.com, rhart@exponent.com, tmorrison@exponent.com

Prepared for Presentation at
American Institute for Chemical Engineers
2023 Spring Meeting
19th Global Congress on Process Safety
Houston, TX
March 12 – 16, 2023

UNPUBLISHED

Blended Natural Gas / Hydrogen Fuel Gas Systems: An Evaluation of Risk

Nicholas Reding, Ph.D., CFEI*
Exponent, Inc.
10850 Richmond Ave., Ste 175, Houston, Texas 66045
nreding@exponent.com

Alfonso Ibarreta, Ph.D., P.E., CFEI; Achim Wechsung, Ph.D., P.E., CFEI;
Ryan Hart, Ph.D., P.E.; Delmar "Trey" Morrison, Ph.D., P.E., CFEI

Keywords: hydrogen fuel gas, blended natural gas, consequence prediction

Abstract

The addition of renewably produced ("green") hydrogen to existing natural gas pipeline systems presents an effective means of transmitting energy with a reduced overall carbon footprint while also granting end-users access to federal monetary incentives, a product of recent climate legislation.

Although the concept of a blended hydrogen/natural gas infrastructure is not new, it presents a host of engineering challenges that merit attention before being put into practice. In consideration of dissimilarities in physical properties and combustion characteristics between natural gas and hydrogen, safe implementation of converted utility systems demands a comprehensive assessment of risks involved, including both probability of accidental leak (infrastructure durability and loss of integrity management, pipeline embrittlement and cracking, and increased leakage rate through permeable elastomer seals) and impact following release of material.

In this paper, a hazard analysis considering the effects of loss of containment for several natural gas/hydrogen blend composition scenarios (0-30% H_2 by volume) is undertaken that specifically examines jet fire thermal radiation, flammable vapor dispersion, and vapor cloud explosions for fixed leak sizes. This investigation applies analytical techniques to evaluate the consequence of potential gas release behavior through vapor cloud dispersion and mixture deflagration simulation. The work aims to identify and quantify the hazard severity and degree of risk associated with system retrofit for hydrogen blending within existing natural gas distribution networks.

* To whom correspondence should be addressed. E-mail: nreding@exponent.com (N. Reding)

1 Introduction

Utilities operate large and complex natural gas transmission and distribution pipeline systems. Recent federal legislation offers clean hydrogen tax benefits for the production of "qualified clean hydrogen" over a 10-year period [1]. At the same time, there is increasing interest from consumers, investors, financial institutions, and regulators to decarbonize. Blending "green" hydrogen (i.e., produced from renewable resources) into natural gas utilities is one opportunity for gas production companies to reduce the carbon footprint associated with their product.

While there is historic precedent for providing consumers with hydrogen-rich gases, today's natural gas transmission and distribution infrastructure has not been developed with hydrogen in mind. Before utilities start flowing hydrogen through existing systems in natural gas service, a host of engineering challenges should be considered. For example, the diffusivity of hydrogen through metallic and polymeric materials differs from those of typical natural gas constituents. Hydrogen may also introduce new mechanical integrity risks due to changes in material durability and increased propensity for steel embrittlement and cracking. Hydrogen has further been known to permeate elastomeric seals and system components at leakage rates dissimilar to natural gas. If risk is defined as a product of incident impact and probability of occurrence, these challenges to implementation constitute an assessment of the likelihood of fuel gas release following conversion of utility service from natural gas to blended hydrogen service. However, when the injected concentration of hydrogen remains low, the adverse effects on infrastructure assets may be avoided such that the incident impact component of risk may be effectively isolated for independent investigation.

In this paper, the results of a hazard analysis are presented. We consider the effects of loss of containment for variable blend ratios of hydrogen (0% - 30% H_2 by volume) and report a comparison of discrete consequences such as jet fire thermal radiation, flammable vapor dispersion, and blast front overpressure for fixed leak sizes. Employing both an integral model as well as a 'first principles' computational fluid dynamics approach, consequence of potential gas release behavior is evaluated through vapor cloud dispersion and mixture deflagration simulation.

2 Hazard Analysis using PHAST

2.1 Methodology

The Process Hazard Analysis Software Tool (PHAST) developed by DNV is a robust integral model, capable of providing an attractive combination of accuracy and low computation cost. Computational hazard analysis using PHAST is split into three stages of calculation, including discharge calculation to describe the initial release from the pipe or vessel rupture, dispersion calculation to describe the behavior of the released material into ambient environment as cloud or pool formation, and effect calculations to describe potential consequence scenarios (toxic exposure, thermal radiation, flash fire, vapor cloud explosion).

Model inputs are exhibited in Table 1. Four fuel gas compositions were investigated, one release of natural gas alone (treated as pure methane for simplicity of calculation) and three separate releases of a natural gas and hydrogen mixture (blends with 10, 20, and 30% hydrogen by volume). Initial pipeline pressure was defined at 800 psi. For each fuel composition of interest, leak diameters of 1, 2, and 4 inches were evaluated. Release duration and elevation were maintained at 600 seconds and 3 feet, respectively, with a horizontal discharge orientation consistently designated. Near ambient weather conditions were utilized for all PHAST simulation scenarios, while also specifying neutral atmospheric stability and 30mm surface roughness.

Scenario (+ xx vol.% H_2)	Natural Gas (NG)	NG + 10% H_2	NG + 20% H_2	NG + 30% H_2
LFL	5.0%	4.9%	4.8%	4.7%
Pressure (psi)	800	800	800	800
Temperature (°F)	68	68	68	68
Phase	Vapor	Vapor	Vapor	Vapor
Leak Diameter (in.)	1, 2, 4	1, 2, 4	1, 2, 4	1, 2, 4
Leak Elevation (ft)	3	3	3	3
Release Orientation	Horizontal	Horizontal	Horizontal	Horizontal
Release Duration (s)	600	600	600	600

Table 1. Release scenarios and input conditions for PHAST analysis.

Distances to various hazards, including jet fire thermal radiation, flammable vapor dispersion, and explosion overpressure, were evaluated with PHAST. Hazard analysis results are presented with respect to the following hazard threshold guidance from API RP 752, *Management of Hazards Associated with Location of Process Plant Permanent Buildings* [2] and API RP 753, *Management of Hazards Associated with Location of Process Plant Portable Buildings* [3]:

- Thermal Radiation: distance to 500, 1,500, 2,000, and 3,000 BTU/ft^2-hr at ground level elevation
- Flammable Vapor Dispersion: distance to lower flammability limit (LFL) and ½-LFL at worst-case flash fire elevation
- Explosion Overpressure: distance to blast front overpressure of 0.6 and 0.9 psi at ground level elevation

2.2 Results and Discussion

2.2.1 Jet Fire Thermal Radiation

Jet fire is a pressurized, or momentum-driven, release of flammable material followed by continuous ignition resulting in turbulent diffusion flame emanating from a source location. If ignition follows an instantaneous release of flammable contents, as in a catastrophic rupture scenario, thermal radiation may take shape as a fireball. Permissible design levels for discrete thermal radiation heat intensities are provided in API Standard 521 [4] and shown in Table 2.

Permissible design level kW/m² (BTU/ft²-hr)	**Conditions**
9.46 (3,000)	Maximum radiant heat intensity at any location where urgent emergency action by personnel is required. When personnel enter or work in an area with the potential for radiant heat intensity greater than 6.31 kW/m² (2,000 BTU/ft²-hr), then radiation shielding and/or special protective apparel (e.g. a fire approach suit) should be considered. SAFETY PRECAUTION – it is important to recognize that personnel with appropriate clothing* cannot tolerate thermal radiation at 6.31 kW/m² (2,000 BTU/ft²-hr) for more than a few seconds.
6.31 (2,000)	Maximum radiate heat intensity in areas where emergency actions lasting up to 30 s can be required by personnel without shielding but with appropriate clothing*
4.73 (1,500)	Maximum radiate heat intensity in areas where emergency actions lasting up to 2-3 minutes can be required by personnel without shielding but with appropriate clothing*
1.58 (500)	Maximum radiate heat intensity at any location where personnel with appropriate clothing* can be continuously exposed.
* Appropriate clothing consists of hard hat, long-sleeved shirts with cuffs buttoned, work gloves, long-legged pants and work shoes. Appropriate clothing minimizes direct skin exposure to thermal radiation.	

Table 2. Permissible design levels for thermal radiant heat intensities.

The PHAST results for jet fire thermal radiation analysis are shown in Figure 1. Blending with hydrogen decreases the distance to any of the thermal radiation hazard thresholds described in Section 2.1, with reductions ranging from 9-12% for a 4" leak of the 30% blend relative to release of natural gas alone. Comparing the analysis report of all simulations, surface emissive power was found to be in a direct relationship with leak size and hydrogen concentration. Radiant energy is proportional to the surface emissive power of the unburnt fuel and combustion products. A larger flame emissive power for hydrogen blends suggests higher radiation at reduced distances and, as a result, tighter proximity of the jet fire hazard thresholds around the initial release location.

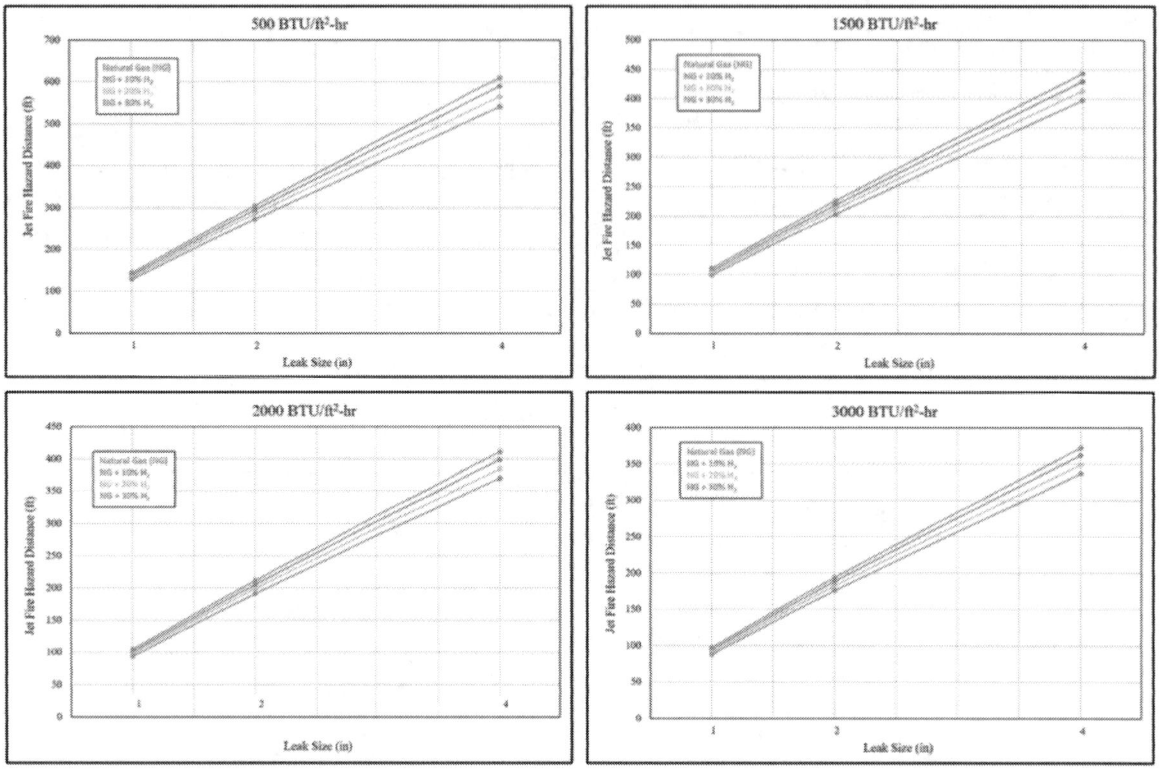

Figure 1. PHAST results for jet fire thermal radiation hazard distances, for thresholds of 500 BTU/ft^2-hr (top left), 1,500 BTU/ft^2-hr (top right), 2,000 BTU/ft^2-hr (bottom left), and 3,000 BTU/ft^2-hr (bottom right).

2.2.2 Flammable Vapor Dispersion

Vapor dispersion involves the release of flammable material, which goes unignited and allows for the mixing with air and formation of a flammable cloud that may be subject to ignition. As flammable material is released, it transitions from jet phase dispersion to heavy and buoyant phase dispersion depending on the specific gravity of mixture components, followed by a transition from momentum-driven active dispersion to passive dispersion. Since the cloud shape fluctuates as the dispersion profile changes, distances to mixture LFL and ½-LFL were measured at worst-case flash fire elevation rather than simply at ground level.

The PHAST results for flammable vapor dispersion analysis are shown in Figure 2.

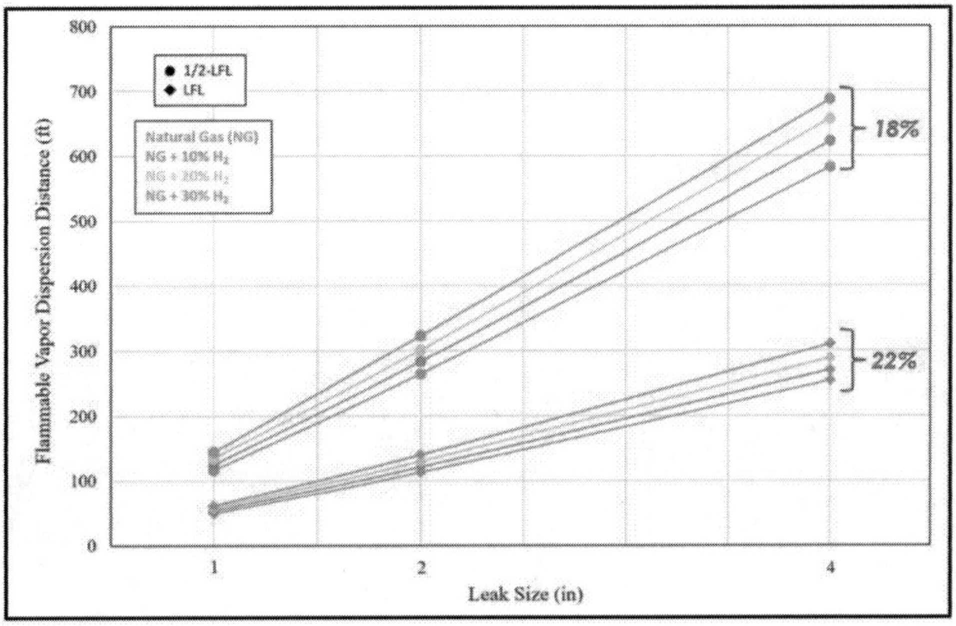

Figure 2. PHAST results for flammable vapor dispersion hazard distances.

As hydrogen concentration from the leak discharge is increased, minimum distance to LFL and ½-LFL increase, with a rise ranging from 17-25% for all leaks sizes for release of the 30% blend relative to natural gas. Methane has a LFL of 5.0% by volume in air; hydrogen gas has a LFL of 4.0% by volume in air. Comparing the mixture LFLs for all hydrogen blends to that of methane, blending with hydrogen reduces the overall mixture LFL, resulting in elongation of the distance from the leak source to the edge of the flammable cloud.

2.2.3 Explosion Overpressure

Overpressure involves pressure effects from blast front propagation following a flammable vapor cloud ignition. In general, five interdependent conditions are required for a vapor cloud explosion to occur, including:

1. Release of flammable material,
2. Sufficient mixing with air to produce a flammable mixture,
3. Delayed ignition which allows the flammable cloud to grow,
4. Ignition of dispersed vapor cloud, and
5. Congestion with or without confinement to accelerate the flame front.

Inherent properties of the material such as flammability range, volatility and dispersion characteristics, and flame propagation behavior influence the severity of a vapor cloud explosion. For the analysis presented in this work, the PHAST simulations applied a blast curve technique known as the Baker-Strehlow-Tang (BST) method to evaluate the blast load from an accelerating flame in a congested area. The PHAST results for explosion overpressure analysis are shown in Figure 3.

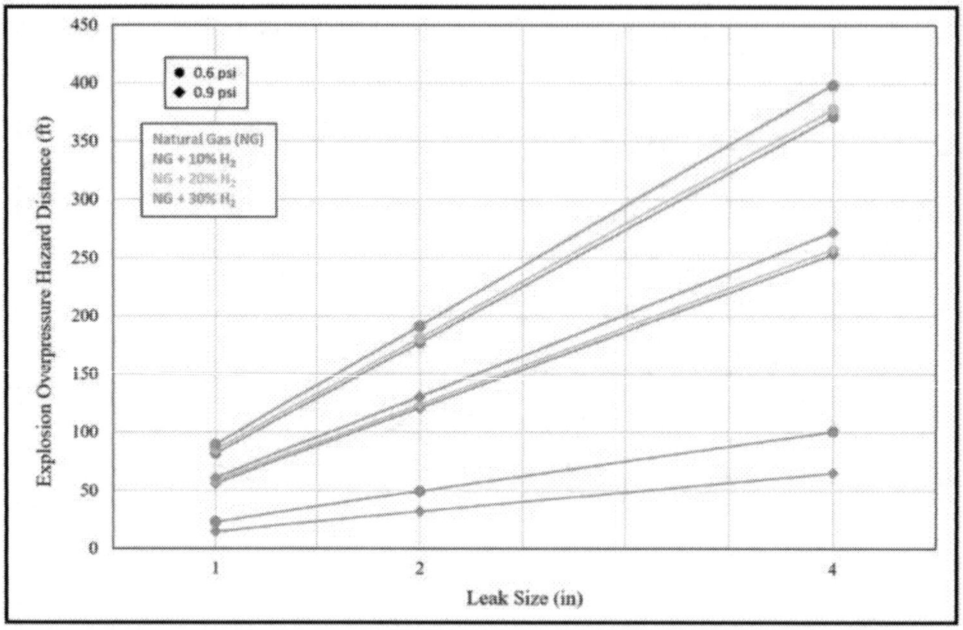

Figure 3. PHAST results for explosion overpressure hazard distances.

For all leak sizes, hydrogen blend release scenarios demonstrated significantly greater overpressure hazard distances relative to release of natural gas. For the 30% blend, explosion overpressure hazard distances increased by an average of 291% and 314% relative to natural gas for the 0.6 psi and 0.9 psi hazard thresholds, respectively. Although the BST model is able to take into consideration the degree of confinement and level of congestion, it requires a relatively broad assignment of fuel reactivity (*low*, *medium*, or *high*). The mixture burning velocity involving fuels of different reactivity, which is treated by summing the individual component burning velocities and is normalized by flammable component mole fraction [5]. Following calculation of mixture fundamental burning velocities, natural gas and the hydrogen blends were assigned as *medium* reactivity fuels (*low* reactivity was assigned for pure natural gas). Such a coarse designation rationalizes the dramatic increase in overpressure hazard distance for any of hydrogen blend scenarios relative to natural gas.

3 Hazard Analysis using CFD

3.1 Methodology

Computation fluid dynamic (CFD) models are able to account for the fuel-air cloud reactivity, as well as the flame front acceleration and fluid flow around complex geometries. CFD models are therefore capable of outputting a refined result tailored to a specific cloud composition, size and geometry configuration at the expense of added computational time. The CFD tool applied for this analysis is the FLACS gas explosion model developed by Gexcon [6]. This software has been validated for use in vapor cloud explosion and confined deflagration modeling. The software spatially resolves the conservation of mass, momentum, and energy on a spatial grid with a sub-model to account for turbulence generation and flame propagation.

Model geometry components for CFD analysis are shown in Figure 4. Congestion objects were created and organized into a 20m x 20m x 20m cubic assembly. A singular congested object was defined using cylinders, where the selected size and spacing provide a 40% area blockage ratio. Congestion is defined as the obstacles that obstruct the passage of the flame front enough to generate turbulence without preventing expansion of the flame front. An increase to the congestion leads to higher turbulent intensity, flame speed, and resultant overpressure.

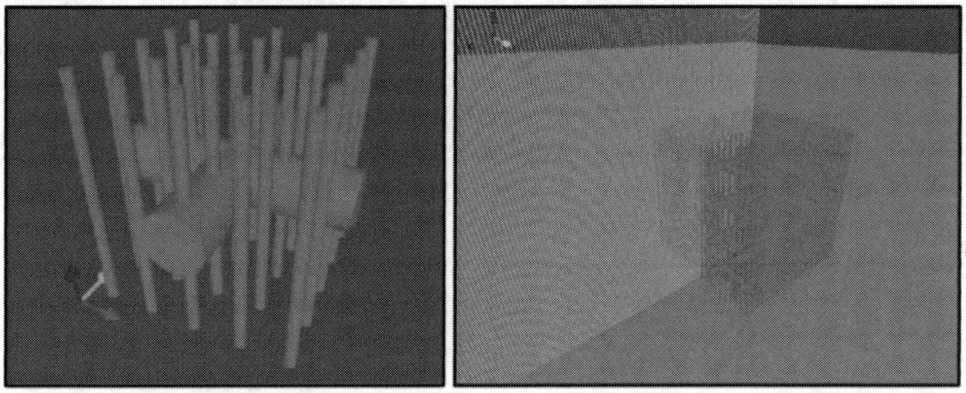

Figure 4. Congestion object (left) and full geometric assembly (right) for CFD analysis.

The total congestion assembly is broken down into four symmetric quadrants. One of the quadrants is simulated, using symmetry conditions on two of the boundaries. The first 5m height of the congested volume is initialized with a fuel gas/air stoichiometric cloud. Assuming stoichiometric combustion in air, the total volume of the initial vapor cloud was determined using the equivalent flammable mass for a 4" release (data extracted from PHAST calculation) and was specified to occupy a corner of the congestion assembly. The location of ignition is set at the very center of the total volume, which is the symmetry corner of the quadrant. The other two walls opposite to the symmetry boundaries have a plane wave boundary condition.

The model domain is 250 m x 250 m x 150 m. The grid sizing is 0.3 m x 0.3 m x 0.3 m within the area of combustion and is stretched to the end of the domain. The pressure is recorded as a function of distance from the geometry center.

3.2 Results and Discussion

For all fuel composition scenarios (natural gas, 10% blend, 20% blend, and 30% blend), explosion pressure was recorded as a function of both time and distance from the congestion assembly. Maximum overpressure, observed at an event time exceeding the extents of the combustion duration, is exhibited in Figure 5.

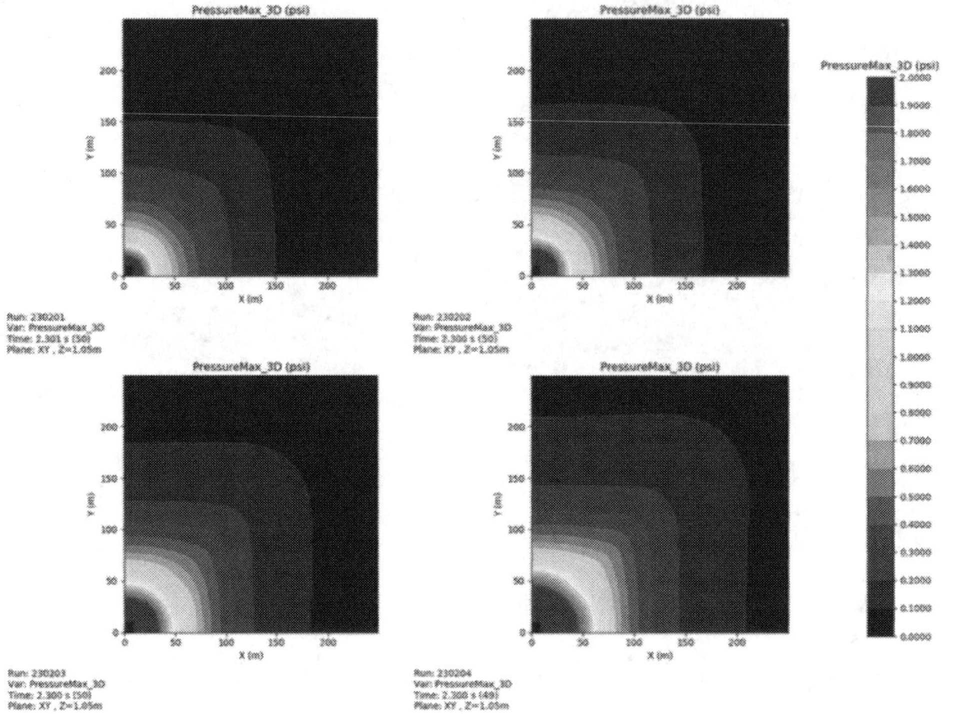

Figure 5. Maximum blast front pressure in the *xy*-directon for all fuel composition scenarios (Natural Gas – top left; 10% blend – top right; 20% blend – bottom left; 30% blend – bottom right).

For the 30% hydrogen blend, the CFD results exhibited explosion overpressure as high as 8 psi at the epicenter. Although the hazard threshold distance evaluated by PHAST remained below the CFD result for release of natural gas, all blending scenarios demonstrated hazard distances consistently larger than those determined using CFD analysis.

The BST model used for PHAST analysis requires an input of fuel reactivity as either *low*, *medium*, or *high*, depending on calculated molar average mixture flame velocity [5]. On the other hand, CFD modeling has potential to apply a more precise turbulent flame speed model for evaluation of blast front overpressure. PHASTs definition of fuel reactivity may be too coarse and speaks to the large jump in hazard distance from pure methane (which was assigned *low* reactivity) to any of the hydrogen blend scenarios (which were all assigned *medium* reactivity). A comparison PHAST and CFD evaluated explosion overpressure results is shown in Figure 6.

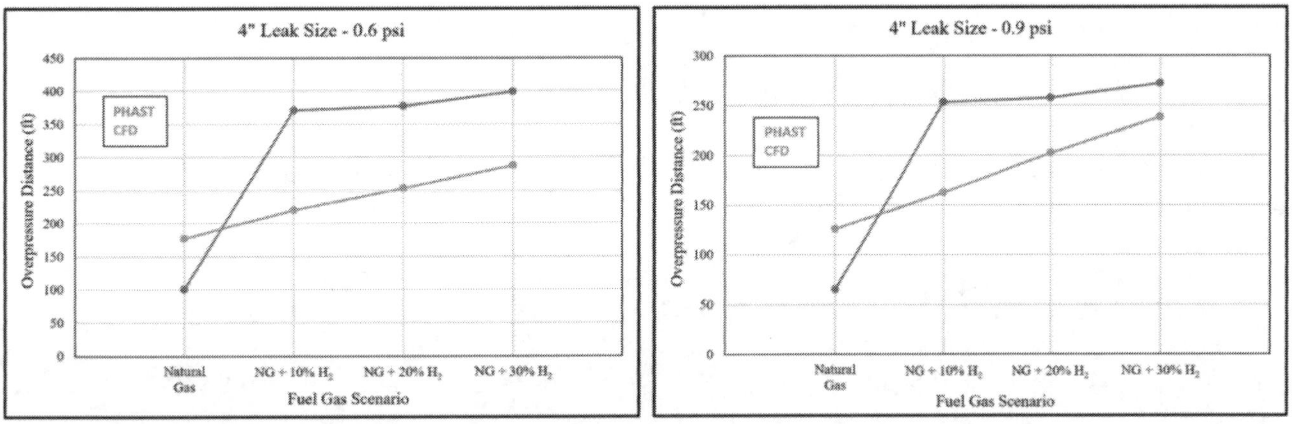

Figure 6. A comparison of PHAST and CFD explosion overpressure results, with distances to both 0.6 psi (left) and 0.9 psi (right) hazard thresholds.

4 Conclusion

In this work, the risk associated with injection of hydrogen to existing natural gas transmission pipelines and distribution networks was evaluated. Applying PHAST and CFD hazard analysis tools, consequence was assessed and contrasted for natural gas and blended hydrogen release scenarios. Hydrogen blends were demonstrated to be a more severe hazard relative to natural gas, with the following findings for the three measured potential consequences:

- **Thermal Radiation**: Reduced proximity jet fire hazard thresholds for hydrogen blends relative to natural gas across all leak sizes
- **Flammable Vapor Dispersion**: A 9-12% increase in flammable vapor hazard distances for release of blended hydrogen relative to natural gas
- **Explosion Overpressure**: Dramatic rise (exceeding 300%, in some cases) of explosion overpressure hazard distances for the release of blends with increased hydrogen content

When a new hazard is introduced, an analysis of impact in the case of accidental release becomes an essential first step toward compliance and safe implementation. As fuel gas production and utility industries look to minimize the risk associated with conversion of existing infrastructure networks from natural gas to hydrogen blending, it is important to expand single-dimension evaluation into a full risk model that incorporates an assessment of both consequence and application-specific probability.

5 References

[1] *Inflation Reduction Act of 2022*, H.R. 5376, 117th Cong. § 13204

[2] Management of Hazards Associated with Location of Process Plant Permanent Buildings, API Recommended Practice 752, 3rd Edition (2009).

[3] Management of Hazards Associated with Location of Process Plant Portable Buildings, API Recommended Practice 753, 1st Edition (2007).

[4] Pressure-Relieving and Depressurizing Systems, API Standard 521, 7th Edition (2020).

[5] Guidelines for Vapor Cloud Explosion, Pressure Vessel Burst, BLEVE and Flash Fire Hazards, 2nd Edition (2010), pp. 196.

[6] Morris, J.; Yen, M.; Ibarreta, A.; Morrison, T.; Hart, R. *Vapor Cloud Explosions in Complex Geometries – Application of the BST Method*, 18th Global Congress on Process Safety (2022).

SAFELY IMPLEMENTING HYDROGEN TO YOUR FACILITY

Maggie Lane, P.Eng., PMP, Ramboll Americas Engineering, Charlotte, NC

Introduction

Hydrogen is a powerful and useful element with properties that can be applied in a vast number of applications. When used safely, there are many uses throughout industry. With the increased focus on hydrogen in a world that is heading toward sustainability, there has been an increasing need in industry to learn about the safety requirements surrounding the implementation of hydrogen systems in existing or new facilities so that the governing codes, standards, and best practices are applied. Some of these codes have undergone significant re-writes and expansions because of these emerging technologies and more significant revisions are expected.

At this stage of industrial development in the US, many companies are ready to introduce hydrogen into their existing processes. For this reason, this paper will be primarily focused on the introduction of hydrogen in an existing facility. Although this is not an exhaustive instruction on how to modify your plant for hydrogen, knowing what questions to ask will go a long way to determining feasibility and/or the modifications needed to existing properties, buildings, HVAC systems, piping, and electrical systems.

Properties of Hydrogen

Understanding the physical properties of hydrogen will be helpful to understanding some of the risks of handling hydrogen and the codes and standards used to design for hydrogen safely.

Gaseous hydrogen is a clear, colorless, and odorless gas in its normal state. With two hydrogen atoms, it is the smallest molecule in the world and is fourteen times lighter than air causing it to be very buoyant in air. In a cryogenic liquid state, the normal boiling point of hydrogen is -423 F (-252.8 C). This is the second lowest boiling point of all the cryogenic liquids next to helium, which has a normal boiling point of -452.1 F (-268.9 C).

The Lower Flammability Limit (LFL) of hydrogen in air is 4% (volume fraction). The Upper Flammability Limit (UFL) is 75%. If the concentration of hydrogen is within 4 and 75% in air, any minor spark could cause a fire. When compared to other common flammable gases in industry as shown in Table 1 below, hydrogen has the second largest flammability range next to Acetylene, which is highly unpredictable and easy to ignite.

Table 1: Flammability limits for common gases used in industry

Common Flammable Gases	LFL (%)	UFL (%)	Range (%)
Acetylene	2.5	80	77.5
Hydrogen	**4.0**	**75.0**	**71.0**
Methane (NG)	5.0	15.0	10.0
Propane	2.1	9.4	7.3
Butane	1.9	8.4	6.5

Main Industrial Hazards Associated with Hydrogen

Flammability

The first and obvious safety hazard concerning hydrogen is flammability. Because of its large flammability range, hydrogen can easily ignite in the presence of air. The ignition of hydrogen is often unpredictable and can be caused by a simple spark from ambient static electricity. In most cases of industrial hydrogen fires, the root cause is never determined.

When hydrogen ignites and burns, it will produce a transparent, light blue flame that is often undetectable by the human eye. When hydrogen ignites, usually due to an accumulation of a cloud, it burns very quickly often in the form of a flash fire. For this reason, Nomex clothing is required near bulk hydrogen storage facilities. While Nomex can burn, it does not ignite easily and protects against flash fires giving the employee time to escape before significant burn injuries can occur.

Asphyxiant

Gaseous hydrogen is a simple asphyxiant at concentrations high enough to dispel air especially as a cold gas when its specific gravity can be close to or higher than that of air. At cold temperatures, hydrogen will tend to hover close to its leak point until it heats up to ambient temperatures and floats above the air.

Freezing and Burns

There is a freezing danger with cryogenic liquid hydrogen. When liquid hydrogen is traveling through pipe, the pipe and piping components will form ice. It is cryogenic ice which forms at temperatures less than -58 F (-50 C). Keeping in mind the temperature of liquid hydrogen, the temperature of the ice accumulating on the pipe is likely around -400F. Therefore, this ice must never be touched without cryogenic gloves or severe, cryogenic burns will occur.

Pressure Hazard

Cryogenic systems are pressurized systems. Therefore, the risk of an unplanned pressure release is inherent to the system. Inside the vessels, the liquid and vapor are in a constant state of change trying to establish equilibrium. Therefore, there are several pressure relief devices installed on a cryogenic liquid vessel to account for any potential upsets that could lead to a pressure spike.

In addition, cryogenic liquids begin to expand to gas as soon as they enter the discharge pipe of the pressure vessel. This expansion quickly increases the pressure in the pipe as the liquid changes state. Therefore, piping designed for cryogenic liquids or gases must have several relief devices specifically designed for the thermal expansion of gas in the pipe in areas where liquid can be trapped. All relief devices in a bulk liquid storage facility are vented to a tall stack.

Liquid Air

An additional hazard is the formation of liquid air on liquid hydrogen piping. When liquid hydrogen is flowing through pipe, a liquid can be seen dripping from the pipe wall or from the ice formed on the pipe. This liquid is often mistaken as moisture (water droplets) from the air. But it is the constituents of air; nitrogen, oxygen and argon liquefying on the pipe. The normal boiling points of these constituents vary from -297 to -320 F.

The hazard with liquid air is that it includes liquid oxygen, which must be kept separate from anything that is easily ignitable. So, liquid hydrogen piping is designed to include metal drip pans to capture the liquid to prevent it from dropping to the ground. Especially, if asphalt or any other petroleum products are present. The liquid air will eventually heat up to ambient conditions and boil off into the atmosphere.

Now that the main hazards of hydrogen systems have been outlined, the codes and standards to safely design for hydrogen can be explored.

Assessing Your Property

If it has been determined that a bulk hydrogen system is required, this equipment is often leased from a hydrogen supplier who will need to assess the property to determine the best location for the installation. The two most common hydrogen storage systems are either a liquid bulk storage system, or a gas bulk storage system.

The ideal location for a bulk storage system must meet certain requirements as outlined by CGA and the hydrogen supplier and may include the following criteria:

1. Adequate accessibility for delivery trucks to enter and exit the property easily and easy access to the bulk storage system by the drivers to fill the tank. Technicians will also require access for maintenance purposes.

2. Adequate security around the property, usually by means of a locked fence. If the property is not fenced in, a fence with gates to the hydrogen vendor's specifications will need to be installed around the bulk storage system with the gates locked.

3. The location of the bulk storage system meets all the minimum separation distances and set-backs from all on-site and off-site exposures as outlined by the National Fire Protection Association (NFPA) and the Compressed Gas Association (CGA). Examples of exposures include:

- Lot lines and parked cars
- Air intakes (HVAC, compressors, etc.)
- Operable openings in building and structures
- Ignition sources such as open flames and welding
- Buildings of combustible construction
- Flammable gas storage systems above or below ground
- Heavy timber, coal, or other slow-burning combustible solids
- Encroachment by overhead utilities
- Piping containing hazardous materials
- Etc.

Once a site location has been chosen, the hydrogen vendor will require and concrete pad to their specified requirements and size be installed on which to install their equipment (assuming the equipment is leased). This is to ensure that liquid air does not drop on any pavement or asphalt and to ensure the stability of the equipment. A grounding system will then be installed to protect the equipment from static charge. Bollards installed every 4 ft around the facility are also required if there is a potential for vehicle damage.

Assessing Your Building

Many industrial hydrogen users do not require a large outdoor bulk storage system. Often users are using hydrogen in small quantities requiring cylinders or cylinder bundles. Inside the building, there are requirements that must be met if cylinders, piping, and equipment using or containing hydrogen are present. The International Building code (IBC), the International Fire Code (IFC), the NFPA and the CGA specify these requirements. The requirements outlined in the IBC/IFC are required by law. IBC/IFC often defer to NFPA and CGA standards and may at times be stricter than NFPA and CGA. Only some of the key requirements will be discussed here. An industrial architect and an HVAC engineer with a firm grasp of flammable gases or vapors should be consulted to assess or design all relevant building features as outline by IBC/IFC and NFPA.

A building may be divided into Control Areas which are usually separated by walls. Each Control Area has a Maximum Allowable Quantity (MAQ) of hydrogen as outlined by the IBC/IFC and re-iterated in several NFPA standards. The MAQ is doubled if the Control Area is sprinklered. If the amount of hydrogen in a control area exceeds the MAQ, the requirements for that area can drastically increase. Such room features to be evaluated include but are not limited to:

- Material and type of construction of the building
- Fire-rating requirements for interior and exterior walls
- What percentage of the walls of the control area must be exterior walls
- Locations of room exhaust vents
- If the MAQ is exceeded, there is a requirement for emergency power for room ventilation in the event of a power failure
- If the MAQ is exceeded, there is a requirement for flammability monitors to be installed to initiate an automatic shutdown of the hydrogen source and activation of an emergency ventilation system if hydrogen is detected in the atmosphere.
- Distance of hydrogen from incompatible materials such as: flammables, pyrophorics, unstable reactives, cryogenics, oxidizers, corrosives, toxics/highly toxics, etc.
- Electrical Area Classifications of areas containing hydrogen

Ventilation

IBC/IFC and NFPA have some significant requirements for room ventilation, especially if the MAQ of hydrogen in a control area is exceeded. Only some key points are highlighted here. Considering that hydrogen gas is fourteen times lighter than air, room exhaust vents must be located within 12 inches of the ceiling with supplemental inlets allowed below 12 inches. One of the most common causes of hydrogen fires is the ignition of hydrogen clouds that was allowed to accumulate near the ceiling of the building.

There are also several requirements on where air intakes can be in relation to ventilation discharge vents and other air handling equipment such as compressors and air conditioning units.

An adequately ventilated Control Area as defined by the IBC/IFC and by NFPA is 1 scf/min/ft^2 of floor area where hydrogen is located. This definition of "adequate ventilation" becomes especially relevant when determining the required electrical classification of the area containing hydrogen.

Electrical Area Classifications

The National Electric Code (NEC) has established a system of electrical classifications for electrical devices that will be used in areas where flammable liquids, gases or vapors are present, as outlined by NFPA. Electrical Area Classifications should be determined either by the hydrogen vendor or by an engineer with a firm grasp of NEC and NFPA electrical area classifications.

Some factors which determine what the electrical classification of an area should be include: Is the area indoors or outdoors? If indoors, is it "adequately ventilated"? Is the vapor or gas lighter than air or heavier that air? Is there an exterior wall or doors nearby? Is there a trench or room exhaust nearby?

While there are several electrical area classifications pertaining to flammables, for the purposes of hydrogen use, the focus will be on areas where flammable vapors or liquids that can form flammable vapors are present:

- Class I, Division 1 (Cl I. Div 1): Areas likely to have an ignitable mixture present under normal operating conditions. "Normal" includes systems that have leaks that should not be there but have not been repaired.
- Class I, Division 2 (Cl I, Div 2): Areas likely to have an ignitable mixture present only under abnormal conditions. "Abnormal" refers to leaks or failures that could release a flammable vapor under abnormal conditions that will be quickly repaired.

Furthermore, chemicals are assigned Groups based on their flammability characteristics. Group A is for Acetylene. Group B is specific to Hydrogen. Most other chemicals are usually either Groups C or D. Since hydrogen has unique flammable properties, it is important to ensure that all classified electrical components in hydrogen service have Group B ratings.

Often, the area within bulk hydrogen storage installation is a Cl I, Div 1 area. This is usually due to the volume of hydrogen present in the storage vessel and there are at least two potential leak sources under normal operations at the vent stacks where hydrogen will vent while maintaining tank equilibrium. In a Class I, Division 1, Group B (Cl I, Div 1, Gr.B) area all electrical instruments, lighting, fixtures, electrical boxes, motors, cell phones or anything that can produce a spark within a certain radius of the installation must have (Cl I, Div 1, Gr.B) ratings.

If all measures have been taken to ensure that there are no leaks inside the building, NFPA allows for the electrical classification of the area to be rated as Class I, Division 2, Group B. Meaning that all electrical components must have Class I, Division 2, Group B (Cl I, Div 2, Gr.B) ratings. More to come on how to design piping to avoid leaks in the next section.

Designing Your Hydrogen Piping Systems

Since most hydrogen piping systems deliver hydrogen at pressures above 15 psig, The American Society of Mechanical Engineers (ASME) Boiler & Pressure Vessel Code (BPVC) applies to process piping and hydrogen delivery systems. More specifically, B31.3 for Process Piping and B31.12 for Hydrogen Piping and Pipelines. The CGA offer best practices and a more comprehensive guide for hydrogen vent stack and supply piping design referencing AMSE B31.12 often. Hydrogen supply piping and vent stacks should be designed by experienced piping engineers that have a firm grasp of the ASME BPVC and experience in designing piping delivery systems for hazardous chemicals.

Piping and Piping Component Materials

The properties of hydrogen offer some unique challenges for piping design in hydrogen service. Because the hydrogen molecule is very small, it can permeate the walls of pipe, filling in the empty spaces between the molecules of the pipe wall. As a result, a phenomenon called "embrittlement" occurs. Piping in cryogenic service, undergoes a constant cycling of expansion and contraction as pipe cools when hydrogen is flowing through it and the heats when hydrogen stops flowing. The embrittlement of metals occurs when hydrogen molecules in the pipe walls expand and contract as the temperature of the pipe fluctuates. Furthermore, the formation of methane molecules within the pores of the metal of the pipe wall causing "chemical embrittlement" can ultimately form cracks in the pipe.

B31.12 addresses both thermal and chemical embrittlement by laying out standards for materials acceptable for use in hydrogen service for industrial piping and for pipeline use. The most common acceptable materials, depending on process conditions, are 300 series seamless austenitic stainless-steel pipe or tube. Other materials can be used can be used under specific conditions and if evaluated for suitability in specific applications. Both ASME and CGA prohibit the use of plastic or PVC piping for use in hydrogen service, except when under strict, low-flow control for lab use.

For stainless steel pipe/tube, the wall thickness will depend on the process conditions. However, wall thicknesses tend to be thicker in hydrogen service than for any other gaseous service to account for embrittlement and to help reduce leaks in non-welded joints.

When specifying piping components such as valves, instruments, inline filters, and other wetted components installed in the pipe, it is important to note that ASME, NFPA and CGA state that such components must be made of materials listed in ASME B31.12. It is also important to note that these components should be rated for hydrogen service. Although yellow metals are rarely used for pipe/tube in hydrogen service, some piping components such as regulators rated for hydrogen service are brass.

Construction of Hydrogen Distribution Piping

Every effort must be taken to avoid the release of hydrogen indoors where hydrogen can form clouds near the ceiling. As a result, joints that can be easily disconnected such as flanged and threaded joints and compression fittings should be avoided as much as possible. Distribution piping/tubing systems in hydrogen service must be welded, inspected, examined and pressure tested per ASME BPVC and B31.12. This will often include 100% non-destructive weld examinations and pressure testing to be performed with gaseous helium, the second smallest molecule in the world next to hydrogen.

There are times when regulators only have NPT options, or when compression fittings or flanges are required at use points. Ways to mitigate leaks at these joints include using the sealant and gasket materials listed as suitable for hydrogen use in ASME B31.12; For compression fittings, using OD tube that is thicker than what would be normally specified for other gases helps to deter the escape of hydrogen. Only double ferrule compression fittings should be used. Single ferrule fittings do not provide a tight enough seal to prevent the escape of hydrogen molecules. Stainless steel OD tube should be installed by experienced pipe fitters that know how to cut, prepare, and clean tube ends for compression fittings to ensure a tight, leak free seal.

Relief Devices

Hydrogen piping must be designed to include thermal relief devices wherever it is possible to trap hydrogen between two closed valves, even when only for maintenance purposes. Pressure relief devices, as in any process piping must be installed where necessary to protect downstream equipment in the event of a pressure regulator failure. All relief devices must be discharged to a vent stack which vents to a safe location outside of the building. The vent stack must be design per CGA guidelines.

Developing Your Training & Safety Procedures

It is important to ensure that clearly established working and safety procedures and lockout / tagout procedures are established when working in or around areas containing hydrogen. Many of the NFPA codes addressing hydrogen reference NFPA 704 which discusses standards for the identification of hazards of materials for emergency response. It includes standards for signage to identify the presence of hydrogen, what the signs should say and where they should be placed.

Once hydrogen has been introduced to your facility, it is important to ensure that all employees are aware of its presence and that they are all trained appropriately according to their roles. Employees that are not directly working on any systems that contain hydrogen should be aware of the basics of hydrogen safety, the electrical classification of the area and what is prohibited in those areas.

Employees such as maintenance personnel and operators working directly on hydrogen systems should be aware of the required PPE and what tools to use. For example, non-sparking tools should be used where hydrogen is present. Non sparking tools are usually made of brass and should include wrenches to replace hydrogen cylinders, screw drivers, etc.

If piping or piping components in hydrogen service require maintenance that require it to be taken out of service, NFPA 56 is the standard for Fire and Explosion Prevention During Cleaning and Purging of Flammable Gas Piping Systems which outlines cleaning and purging of hydrogen systems before maintenance work can begin, especially if the work requires hot work such as welding.

Health and Safety personnel should be aware of all the hazards of hydrogen and what emergency response and first aid procedures need to be developed and implemented.

The best resource to turn to for help to develop processes and procedures aside from the listed codes and standards are often the hydrogen supplier. They have well established standard operating procedures and safety procedures and often have staff that can train key personnel at your facility.

Conclusions

The purpose of this paper was to increase the level of comfort when using hydrogen by discussing the design standards and requirements that need to be considered when hydrogen is being introduced to an existing or a new facility. Hopefully, readers are more comfortable with the idea of hydrogen at their facility and know what questions to ask at the beginning of their project so that there are no surprises that can affect budgets or schedules later.

References

1. International Building Code with amendments
2. International Fire Code with amendments
3. ASME B31.12 Hydrogen Piping and Pipelines
4. CGA G-5.5 Hydrogen Vent Systems
5. CGA G-5.4 Standard for Hydrogen Piping Systems at User Locations
6. CGA H-5 Standard for Bulk Hydrogen Supply Systems
7. NFPA 2 Hydrogen Technologies Code
8. NFPA 55 Compressed Gases and Cryogenic Fluids Code
9. NFPA 56 Standard for Fire and Explosion Prevention During Cleaning and Purging of Flammable Gas Piping Systems
10. NFPA 70 National Electric Code
11. NFPA 497 Recommended Practice for the Classification of Flammable Liquids, Gases or Vapors and of Hazardous (Classified) Locations for Electrical Installations in Chemical Process Areas
12. NFPA 704 Standard System for the Identification of the Hazards of Materials for Emergency Response

AUTHOR INDEX

AIChE
120 Wall Street, FL 23
New York, NY 10005-4020

ISBN 978-1-7138-7936-7

9 781713 879367

CURRAN ASSOCIATES INC.
proceedings.com